PLANT GATEWAY'S

THE GLOBAL FLORA

A practical flora to vascular plant species of the world

INTRODUCTION

Introducing *The Global Flora*
The phylogeny of angiosperms poster

January 2018

The Global Flora

A practical flora to vascular plant species of the world

Introduction, Vol. 1: 1-35.

Published by Plant Gateway Ltd., 5 Baddeley Gardens, Bradford, BD10 8JL, United Kingdom

ISSN   2398-6336
eISSN 2398-6344
ISBN 978-1-912629-00-8
eISBN 978-0-9929993-9-1

Plant Gateway has no responsibility for the persistence or accuracy of URLS for external or third-party internet websites referred to in this work, and does not guarantee that any content on such websites is, or will remain, accurate or appropriate.

British Library Cataloguing in Publication data
A Catalogue record of this book is available from the British Library

For information or to purchase other Plant Gateway titles please visit
www.plantgateway.com

Cover image: Simplified angiosperm phylogeny © James Byng

Introducing *The Global Flora,* a global series of botany

James W. Byng[1,2] & Maarten M.J. Christenhusz[1,3]

[1]*Plant Gateway Ltd., 5 Baddeley Gardens, Bradford, BD10 8JL, UK.*
[2]*Naturalis Biodiversity Center, P.O. Box 9517, 2300 RA Leiden, the Netherlands.*
[3]*Royal Botanic Gardens, Kew, Richmond, Surrey TW9 3DS, UK.*

Species Plantarum by Linnaeus (1753) contained 5,940 species of plants, including all known species then known globally. Since its publication 264 years ago, the exploration of plant diversity across the planet has led to approximately 374,000 known, described and accepted plant species (Christenhusz & Byng, 2016). This number increases by approximately 2000 additional new species each year, despite the slow but steady decrease in the number of active herbarium taxonomists focused on monography. Current estimates suggest that at least twenty percent of all botanical diversity still remains to be discovered, analysed (or re-analysed), described and named.

Herbaria are exceptionally rich sources of botanical information. They house voucher specimens used for anatomical, biogeographical, chemical, molecular, morphological, palynological and taxonomical studies. Type specimens, which permanently link a scientific name to a physical specimen, effectively are the "birth certificates" of each species, and are amongst the most important specimens in each herbarium. Herbarium specimens collectively provide critical baseline data for where and when a species occurred, and they can be used to evaluate population increases or declines. Thus herbarium collections provide the physical evidence for much of our botanical knowledge.

An ongoing and acute problem in taxonomy is that large numbers of specimens in nearly all herbaria are either unidentified or identified incorrectly (e.g. Bebber *et al.*, 2010; Goodwin *et al.*, 2015). This does not reflect incompetence or negligence on the part of botanists, but rather a dearth of active taxonomists (Uno, 2009) and low (and declining) levels of funding for basic research (Dalton *et al.*, 2003; Agnarsson & Kuntner, 2007; Ahrends *et al.*, 2010). Taxonomic studies are never-ending as new species are described, new specimens are collected and need identifying and older treatments need to be updated following new findings in nomenclature, taxonomy and evolutionary botany.

Fortunately, some positive developments are occurring in systematics given the advent of digitial technologies. The recent acceptance of electronic publication in the *International Code of Nomenclature for Algae, Fungi and Plants* has significantly expedited species discovery and the solving of associated nomenclatural issues (McNeill *et al.*, 2012; Christenhusz & Byng, 2016). But while digital technologies and electronic publication are helping to speed up some aspects of taxonomy, fewer treatments having a global taxonomic focus are being produced. Those who must rely heavily on these treatments, such as ecologists, conservation biologists, and invasive species biologists often are fully aware that no reliable and current source of current taxonomic information exists at the global level for many groups of plants.

Plant Gateway is working on various higher-level classifications and overviews of vascular plants (e.g. Christenhusz *et al.*, 2011; Christenhusz & Chase, 2014; Byng, 2014, 2015; APG IV, 2016; Christenhusz & Byng 2016; Christenhusz *et al.*, 2017; Byng *et al.*, 2018). In this framework, we are introducing *The Global Flora*, a new international serial for botanical taxonomy, to provide accepted species-level classifications for all vascular plant families based on available or generated molecular data and re-examining the literature and herbarium specimens in major herbaria. The goal is to provide a current, balanced and practical taxonomy reflecting evolutionary relationships.

The Global Flora will be available in print and online versions and include three series: (A) Angiosperms (following APG IV, 2016); (B) Lycopods, Ferns and Gymnosperms (classification following Christenhusz *et al.*, 2017); and (C) special editions. The first two series will only treat monophyletic taxa on a global scale (e.g. family, subfamily, tribe, genus or section). The content and format of each taxonomic treatment in the first two series will vary depending on the group, and interested authors should consult the first few published treatments for guidance. The special editions series aims to make significant contributions to the body of plant systematic knowledge and typically will be of a global botanical scope.

The Global Flora will be published frequently and at regular intervals. It will be amply illustrated and should appeal to many different users, including ecologists, conservationists, gardeners and other plant enthusiasts in the applied sciences, as opposed to appealing solely to practicing taxonomists. As important new evidence becomes available updates and revisions to already published treatments will be allowed to make the treatments current and dynamic. Unlike other journals and flora serials, *The Global Flora* will share royalties with authors and compensate reviewers and editors for their respective duties. This is important for maintaining momentum, because relatively few institutions provide time and funds for researchers to do this time-consuming, yet crucially important work. An unspoken consensus of many taxonomists is that the historical lack of remuneration for this type of research is a primary reason that taxonomy does not progress more rapidly.

In addition, we believe that some of the funds generated from taxonomic work should be retained within the taxonomic community. For reviewers and contributors who are unable to accept royalties as part of their positions, or who choose to waive them altogether, the royalties will be transferred to a *Global Flora Small Grants Fund* to allow future contributors to undertake herbarium visits and generate data for future treatments in *The Global Flora*. In return for a fee, authors may opt for their taxonomic treatments to be open-access, which will cover editorial costs, and the remainder of which shall be added to the *Global Flora Small Grants Fund*. All special edition issues will be open-access.

We hope our readers enjoy and support *The Global Flora* in its initial stages. An Editorial Board is being formed as our first issues appear. We also hope that the contents of *The Global Flora* will provide significant contributions to help achieve Target 1 of the Global Strategy for Plant Conservation established for 2020, and we hope the series will grow towards a global and complete coverage of the vascular plants of the world.

References

Agnarsson, I. & Kuntner, M. (2007) Taxonomy in a changing world: seeking solutions for a science in crisis. *Systematic Biology* 56: 531–539.

Ahrends, A., Rahbek, C., Bulling, T.M., Burgess, N.D., Platts, P.J., Lovett, J.C., Kindemba, V.W., Owen, N., Sallu, A.N., Marshall, A.R., Mhoro, B.E., Fanning, E. & Marchant, R. (2010) Conservation and the botanist effect. *Biological Conservation* 144: 131–140.

APG IV. (2016) An update of the Angiosperm Phylogeny Group classification for the orders and families of flowering plants: APG IV. *Botanical Journal of the Linnean Society* 181: 1–20.

Bebber, D.P., Carine, M.A., Wood, J.R.I., Wortley, A.H., Harris, D.J., Prance, G.T., Davidse, G., Paige, J., Pennington, T.D., Robson, N.K.B. & Scotland, R.W. (2010) Herbaria are a major frontier for species discovery. *Proceedings of the National Academy of Sciences of the United States of America* 107: 22169–22171.

Byng, J. W. (2015) *The Gymnosperms Handbook: A practical guide to extant families and genera of the world*. Plant Gateway: Hertford. 36 pp.

Byng, J. W. (2014[2015]) *The Flowering Plants Handbook: A practical guide to families and genera of the world*. Plant Gateway: Hertford. 619 pp.

Byng, J.W., Smets, E., Vugt, R. van, Bidault, E., Davidson, C., Kenicer, G., Chase, M.W. & Christenhusz, M.J.M. (2018) The phylogeny of angiosperms poster. *The Global Flora* 1: 4–35.

Christenhusz, M.J.M. & Byng, J.W. (2016) The number of known plant species in the world and its annual increase. *Phytotaxa* 261: 201–217.

Christenhusz, M.J.M. & Chase, M.W. (2014) Trends and concepts in fern classifications. *Annals of Botany* 113: 571–59.

Christenhusz, M.J.M., Fay, M.F. & Chase, M.W. (2017) *Plants of the World. An illustrated encyclopedia of vascular plants*. Kew Publishing: Richmond & Chicago University Press: Chicago. 792 pp.

Christenhusz, M. J. M., Chase, M. W. & Fay, M. F. (eds.). (2011) Linear sequence, classification, synonymym and bibliography of vascular plants. *Phytotaxa* 19: 4-134.

Dalton, R. (2003) Natural history collections in crisis as funding is slashed. *Nature* 423(6940): 575–575.

Goodwin, Z.A., Harris, D.J., Filer, D., Wood, J.R. & Scotland, R.W. (2015) Widespread mistaken identity in tropical plant collections. *Current Biology* 25(22): R1066–R1067.

Linnaeus, C. (1753) *Species plantarum, exhibentes plantas rite cognitas ad genera relatas, cum differentiis specificis, nominibus trivialibus, synonymis selectis, locis natalibus, secundum systema sexuale digestas*. Laurentius Salvius: Stockholm, 1200 pp.

McNeill, J. Barrie, F.R., Buck, W.R., Demoulin, V., Greuter, W., Hawksworth, D.L., Herendeen, P.S., Knapp, S., Marhold, K., Prado, J., Prud'homme van Reine, W.F., Smith, G.F., Wiersema, J.H. & Turland, N.J. (2012) *International Code of Nomenclature for Algae, Fungi and Plants (Melbourne Code)*. Koeltz: Koenigstein.

Uno, G. (2009) Botanical literacy: What and how should students learn about plants? *American Journal of Botany* 96: 1753–1759.

About the Chief-Editors

Maarten M.J. Christenhusz BSc MSc PhD

London, United Kingdom

m.christenhusz@kew.org

A Dutch born botanist who is deputy chief-editor of the *Botanical Journal of the Linnean Society* and was the founding chief-editor of the international taxonomic journal *Phytotaxa*. He completed an MSc in Biology from Utrecht University (Netherlands) and a PhD in Biodiversity from Turku University (Finland) and has worked at the Natural History Museum in London, The Finnish Museum of Natural History in Helsinki and the Royal Botanic Gardens, Kew. He is president of the International Pteridological Society, a member of the Society of Authors and a fellow of the Linnean Society of London. He has authored numerous papers, flora treatments and books on vascular plant taxonomy and was one of the compilers of the latest Angiosperm Phylogeny Group classification (APG IV, 2016). He is also the curator of Plant Gateway's teaching herbarium (PG).

James W. Byng BSc (Hons.) MSc PhD

Leiden, the Netherlands

james.byng@naturalis.nl

A British born botanist who is the founder of Plant Gateway Ltd. and a Research Fellow at Naturalis Biodiversity Center in Leiden (Netherlands). He completed his PhD at the University of Aberdeen (UK) and Royal Botanic Gardens, Kew (UK) and previously completed the MSc in Biodiversity and Taxonomy of Plants at the University of Edinburgh and Royal Botanical Gardens, Edinburgh (UK). He has authored numerous papers, flora treatments and books on vascular plant taxonomy and was one of the compilers of the latest Angiosperm Phylogeny Group classification (APG IV, 2016). He is also a subject editor for the online journal *Phytokeys* and is a fellow of the Linnean Society of London.

The phylogeny of angiosperms poster: a visual summary of APG IV family relationships and floral diversity

James W. Byng[1,2], Erik F. Smets[2,3], Rogier van Vugt[4], Ehoarn Bidault[5,6], Christopher Davidson[7], Greg Kenicer[8], Mark W. Chase[9,10] & Maarten J.M. Christenhusz[1,9, 11]

[1]Plant Gateway, 5 Baddeley Gardens, Bradford, BD10 8JL, UK.

[2]Naturalis Biodiversity Center, P.O. Box 9517, 2300 RA Leiden, The Netherlands.

[3]Section Ecology, Evolution and Biodiversity Conservation, KU Leuven, BE-3001 Leuven, Belgium.

[4]Hortus Botanicus of Leiden University, Rapenburg 73, 2311 GJ Leiden, The Netherlands.

[5]Missouri Botanical Garden, Africa & Madagascar Department, P.O. Box 299, St. Louis, Missouri 63166-0299, USA.

[6]Institut de Systématique, Evolution, et Biodiversité ISYEB UMR 7205 CNRS/ MNHN/ EPHE/ UPMC, Muséum national d'Histoire naturelle, Sorbonne Universités, C.P.39, 57 rue Cuvier, F-75231 Paris CEDEX 05, France.

[7]Idaho Botanical Research Foundation, 637 Warm Springs Ave., Boise, Idaho 83712, U.S.A.

[8]Royal Botanic Garden Edinburgh, Edinburgh EH3 5LR, UK.

[9]Royal Botanic Gardens, Kew, Richmond, Surrey TW9 3DS, UK.

[10]University of Western Australia, 35 Stirling Highway, Crawley, Western Australia 6009, Australia.

[11]Curtin University, GPO Box U1987, Perth, Western Australia 6845, Australia.

Abstract

This article provides a visual overview of the relationships of all angiosperm families (following APG IV). The poster lists important characters for major grades and clades and these are illustrated with flower images of 269 plant families. It is presented to provide a useful educational tool. The scientific names and photo accreditation of each image are listed.

Keywords

Angiosperms – Flower images – Phylogeny – Poster – Taxonomy – Teaching

Background

It is widely accepted in many countries that courses, practicals and lectures focusing on plant taxonomy have been in decline during the last few decades (e.g. Hershey, 1996; Woodland, 2007; Drea, 2011; Uno, 2011). Many undergraduate biology students find currently that learning basic classification and relationships of plants are taught often during a single practical, which results in many young biologists graduating with little understanding of botanical diversity and without having sufficient identification and taxonomic skills. Current students know far less about plant family relationships than before, despite scientific insights being better.

There has been a molecular revolution in the field of plant taxonomy during the last 25 years, where thousands of phylogenetic trees have been generated using molecular sequences from thousands of plant species. Our understanding of higher-level classification of angiosperms have improved substantially, which have led to four versions of the widely used and primarily molecular based Angiosperm Phylogeny Group (APG) classification (APG 1998; APG II 2003; APG III 2009; APG IV 2016). Unexpected relationships,

like that of Nelumbonaceae with Proteaceae, were uncovered and circumscriptions for several families (e.g. Malvaceae, Molluginaceae, Portulacaceae, Salicaceae, Sapindaceae, Scrophulariaceae) have substantially changed with molecular data and this often makes much of the available literature difficult to use for teaching, although several new works aiding in the teaching of plant taxonomy, identification and economic botany have recently been published following the modern classifications like APG (e.g. Byng, 2014; Christenhusz et al., 2017). However, many of the plant taxonomic works used for teaching are too technical for beginners or too regionally focused. Because usually only a small subset of plant families is taught students get confused or despair in the greater diversity they may discover when they visit gardens or greenhouses where cultivated plants are grown from all around the world.

In The phylogeny of angiosperms poster (Figure 1) we present a global overview of the relationships of all 416 families in the 64 orders following the APG IV classification (Figures 1, 2, 3, 5, 8, 10, 12, 14, 16, 19, 21, 23, 26). Phylogenetic trees that include all APG IV families are few, because the APG classifications were based on a number of independent papers that treat parts of the angiosperms, and the one presented in the poster is discussed in more detail in Byng *et al.* (in prep.), notably concerning the orders Ceratophyllales and Dilleniales.

The poster includes 269 flower images (Figures 1, 4, 6, 7, 9, 11, 13, 15, 17, 18, 20, 22, 24, 25, 27) showing a diverse range of floral morphology. We aim to show this diversity to enthuse students and teachers and anyone with an interest in plants. It provides a visually stunning tool for education and it will be a practical resource for teaching introductory and advanced classes in botany. We hope it will inspire students to pursue a future career in botany.

Acknowledgements

We thank the many photo contributors who have allowed us to use images. Also, we thank Stephan Eckel, Florian Jabbour, Farah Rahman, Neil Snow and Gerda van Uffelen for comments on early drafts of the poster.

References

APG (Angiosperm Phylogeny Group). (1998) An ordinal classification for the families of flowering plants. *Annals of the Missouri Botanical Garden* 85: 531–553.

APG II. (2003) An update of the Angiosperm Phylogeny Group classification for the orders and families of flowering plants: APG II. *Botanical Journal of the Linnean Society* 141: 399–436.

APG III. (2009) An update of the Angiosperm Phylogeny Group classification for the orders and families of flowering plants. APG III. *Botanical Journal of the Linnean Society* 161: 105–121.

APG IV. (2016) An update of the Angiosperm Phylogeny Group classification for the orders and families of flowering plants: APG IV. *Botanical Journal of the Linnean Society* 181: 1–20.

Byng, J.W. (2017).

Byng, J.W., Smets, E., Chase, M.W. & Christenhusz, M.J.M. (in prep.) The phylogeny of angiosperms: an overview of current relationships. *The Global Flora*.

Christenhusz, M.J.M., Fay, M.F. & Chase, M.W. (2017). *Plants of the World, an illustrated encyclopedia to vascular plant families*. Kew Publishing, Richmond.

Drea, S. (2011) The end of the botany degree in the UK. *Bioscience education* 17: 1–7.

Hershey, D.R. (1996) A historical perspective on problems in botany teaching. *The American Biology Teacher* 58: 340–347.

Uno, G.E. (2009) Botanical literacy: what and how should students learn about plants? *American Journal of Botany* 96: 1753–1759.

Woodland, D.W. (2007) Are botanists becoming the dinosaurs of biology in the 21st century? *South African Journal of Botany* 73: 343–346.

Figure 1: Overview of The phylogeny of angiosperms poster.
Full download available from: http://www.plantgateway.com/poster/

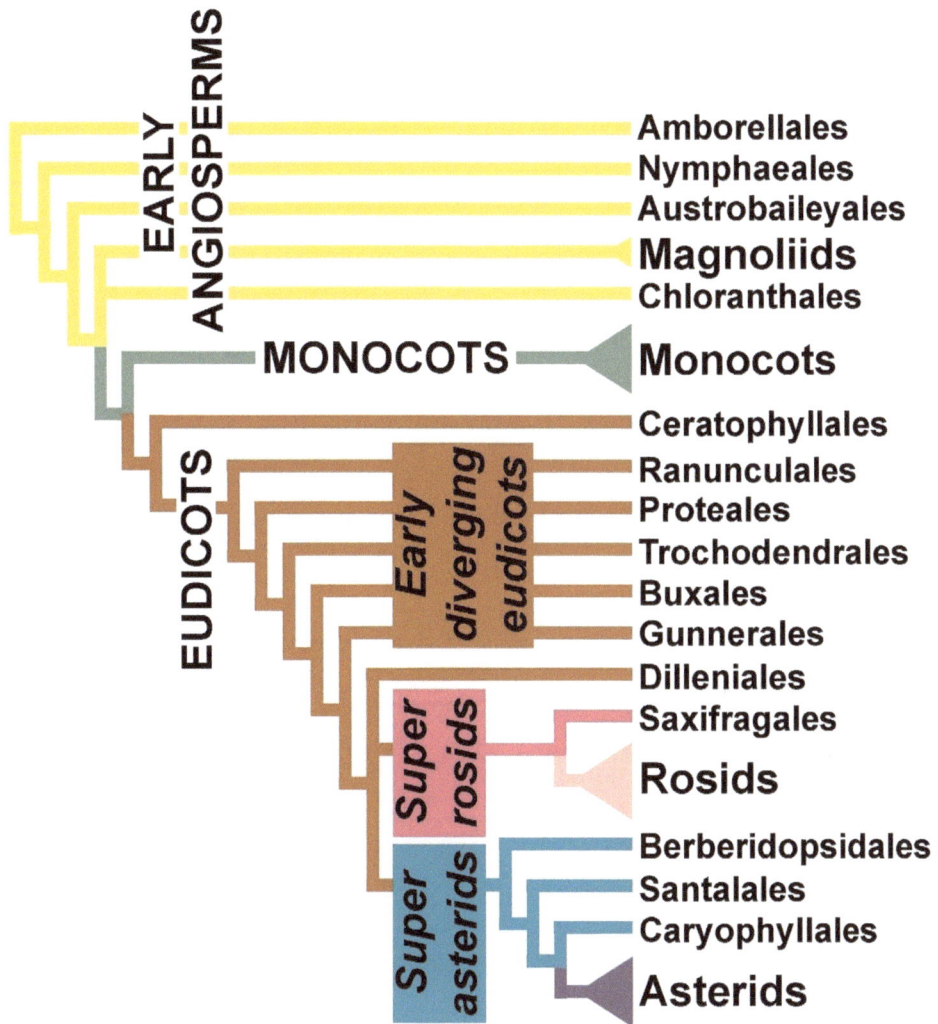

Figure 2: A simplified phylogeny of angiosperms.

Early angiosperms

8 orders : 26 families
(ANA-grade + magnoliids + Chloranthales)

Two cotyledons almost always present

Ethereal oils often present

Leaves almost always simple, net-veined

Usually many floral parts to each whorl

Perianth usually spiralling or parts in threes

Stamen filaments usually broad

Anthers tetrasporangiate

Pollen monosulcate

Nectaries rare

Carpels usually free

Embryo very small

Amborellales — 1. Amborellaceae

Nymphaeales
- 2. Hydatellaceae
- 3. Cabombaceae
- 4. Nymphaeaceae

Austrobaileyales
- 5. Austrobaileyaceae
- 6. Trimeniaceae
- 7. Schisandraceae

Canellales
- 8. Canellaceae
- 9 . Winteraceae

Piperales
- 10. Saururaceae
- 11. Piperaceae
- 12. Aristolochiaceae

Magnoliales
- 13. Myristicaceae
- 14. Magnoliaceae
- 15. Degeneriaceae
- 16. Himantandraceae
- 17. Eupomatiaceae
- 18. Annonaceae

Laurales
- 19. Calycanthaceae
- 20. Siparunaceae
- 21. Gomortegaceae
- 22. Atherospermataceae
- 23. Hernandiaceae
- 24. Monimiaceae
- 25. Lauraceae

Chloranthales — 26. Chloranthaceae

MAGNOLIIDS

OTHER ANGIOSPERMS

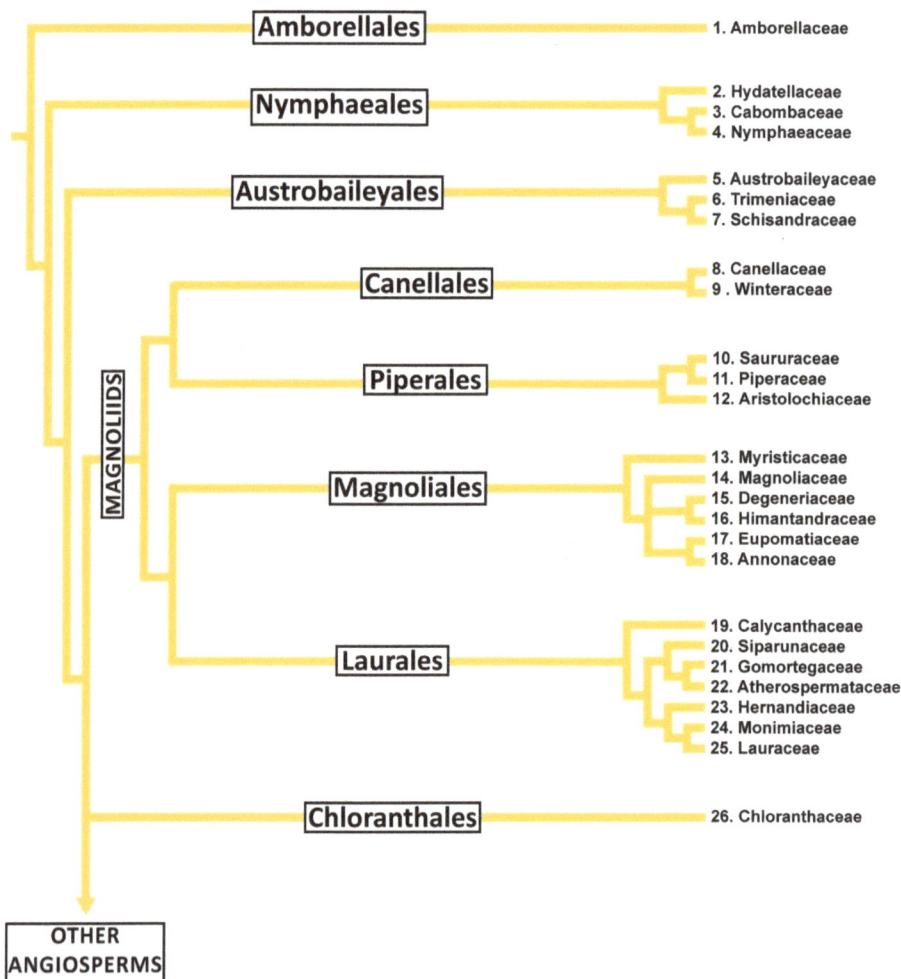

Figure 3: Diagnostic characters and relationships of early angiosperms.

Figure 4: Floral images of Amborellales (family 1), Nymphaeales (2-4), Austrobaileyales (5-7), Canellales (9), Piperales (10-12), Magnoliales (13-18), Laurales (19-25) and Chloranthales (26).

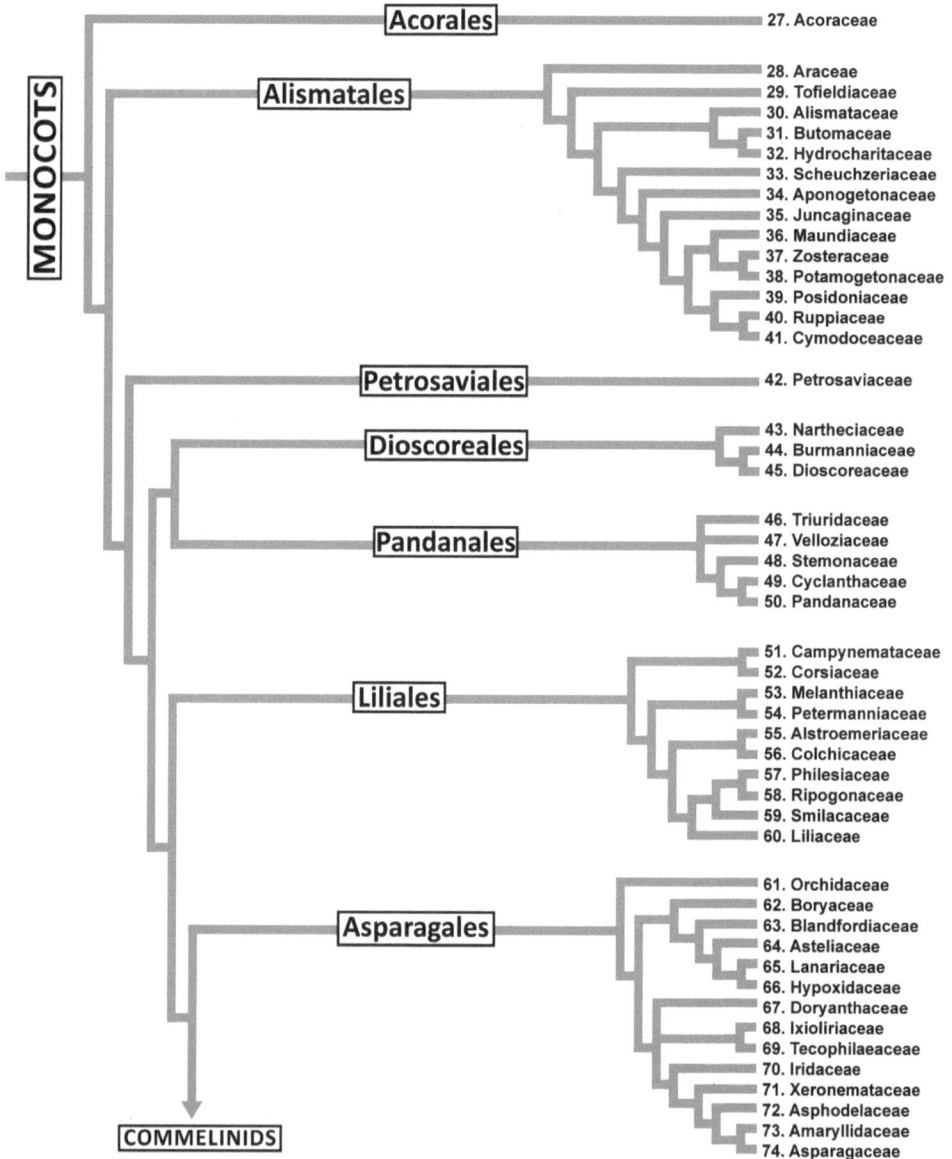

Monocots

11 orders : 77 families
(basal monocots + lilioid + commelinids)

Single cotyledon present

Usually a short-lived primary root

Single adaxial prophyll

Ethereal oils rarely present

Mostly herbaceous, vascular cambium absent

Scattered vascular bundles in the stem

Leaves simple, usually parallel-veined.

Floral parts usually in threes

Perianth often composed of tepals

Pollen monosulcate

Styles frequently hollow

Successive microsporogenesis

MONOCOTS

Acorales
- 27. Acoraceae

Alismatales
- 28. Araceae
- 29. Tofieldiaceae
- 30. Alismataceae
- 31. Butomaceae
- 32. Hydrocharitaceae
- 33. Scheuchzeriaceae
- 34. Aponogetonaceae
- 35. Juncaginaceae
- 36. Maundiaceae
- 37. Zosteraceae
- 38. Potamogetonaceae
- 39. Posidoniaceae
- 40. Ruppiaceae
- 41. Cymodoceaceae

Petrosaviales
- 42. Petrosaviaceae

Dioscoreales
- 43. Nartheciaceae
- 44. Burmanniaceae
- 45. Dioscoreaceae

Pandanales
- 46. Triuridaceae
- 47. Velloziaceae
- 48. Stemonaceae
- 49. Cyclanthaceae
- 50. Pandanaceae

Liliales
- 51. Campynemataceae
- 52. Corsiaceae
- 53. Melanthiaceae
- 54. Petermanniaceae
- 55. Alstroemeriaceae
- 56. Colchicaceae
- 57. Philesiaceae
- 58. Ripogonaceae
- 59. Smilacaceae
- 60. Liliaceae

Asparagales
- 61. Orchidaceae
- 62. Boryaceae
- 63. Blandfordiaceae
- 64. Asteliaceae
- 65. Lanariaceae
- 66. Hypoxidaceae
- 67. Doryanthaceae
- 68. Ixioliriaceae
- 69. Tecophilaeaceae
- 70. Iridaceae
- 71. Xeronemataceae
- 72. Asphodelaceae
- 73. Amaryllidaceae
- 74. Asparagaceae

COMMELINIDS

Figure 5: Diagnostic characters and relationships of monocots.

Figure 6: Floral images of Acorales (family 27), Alismatales (28-38), Petrosaviales (42), Dioscoreales (44-45), Pandanales (46-50) and Liliales (52-57).

Figure 7: Floral images of Liliales (families 59-60) and Asparagales (61-74).

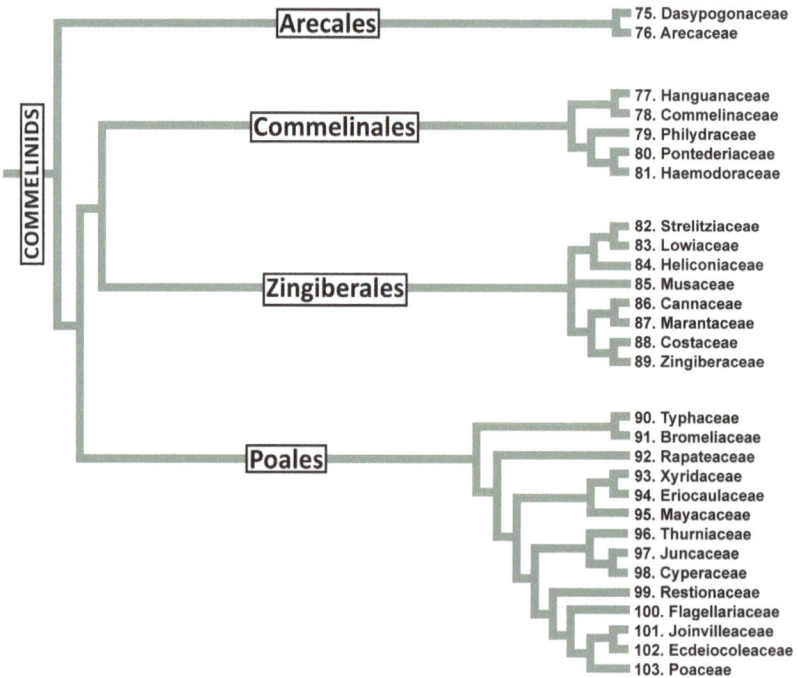

Figure 8: Relationships of commelinid monocots.

Figure 9: Floral images of the Arecales (family 76), Commelinales (78-81), Zingiberales (82-89) and Poales (90-103).

Eudicots

45 orders : 313 families
(early diverging eudicots + superrosids + superasterids)

Two cotyledons almost always present	Plants woody or herbaceous	Styles usually solid
Nodes trilacunar with three leaf traces	Leaves simple or compound, usually net-veined	Pollen tricolpate
Stomata anomocytic	Flower parts mostly in twos, fours or fives	
Ethereal oils rarely present	Microsporogenesis simultaneous	

Early diverging eudicots   7 orders : 17 families
+ Ceratophyllales & Dilleniales

Morphological transition grade

Exhibit ancestral and derived characters

Pollen tricolpate

Sometimes flowers with many stamens, free carpels and flower parts in threes or many parts like *Early angiosperms*

Sometimes flower parts in fours or fives like many *Core eudicots*

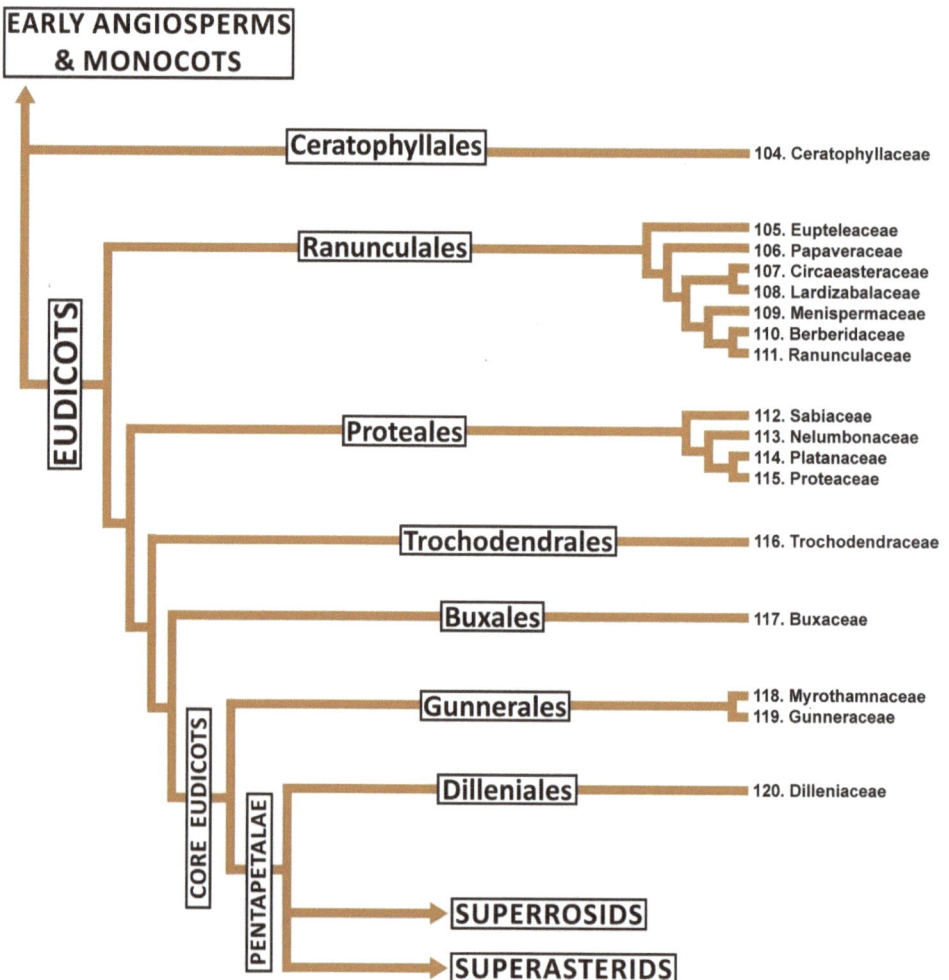

Figure 10: Diagnostic characters and relationships of early diverging eudicots.

Figure 11: Floral images of Ranunculales (families 106-111), Proteales (112-115), Trochodendrales (116), Buxales (117), Gunnerales (119) and Dilleniales (120).

Superrosids

**18 orders : 150 families
(Saxifragales + rosids)**

Rosids

**17 orders : 135 families
(Vitales + fabids + malvids)**

Stipules often present (about 65% of families)

Petals usually free (about 97%)

Receptacular nectaries and hypanthia are common

Stamens often in 2 whorls to many (about 58%)

Thick (crassinucellate) nucellus (about 92%)

Two integuments covering the ovule (about 95%)

Endosperm nuclear (about 95%)

Iridoids rarely present

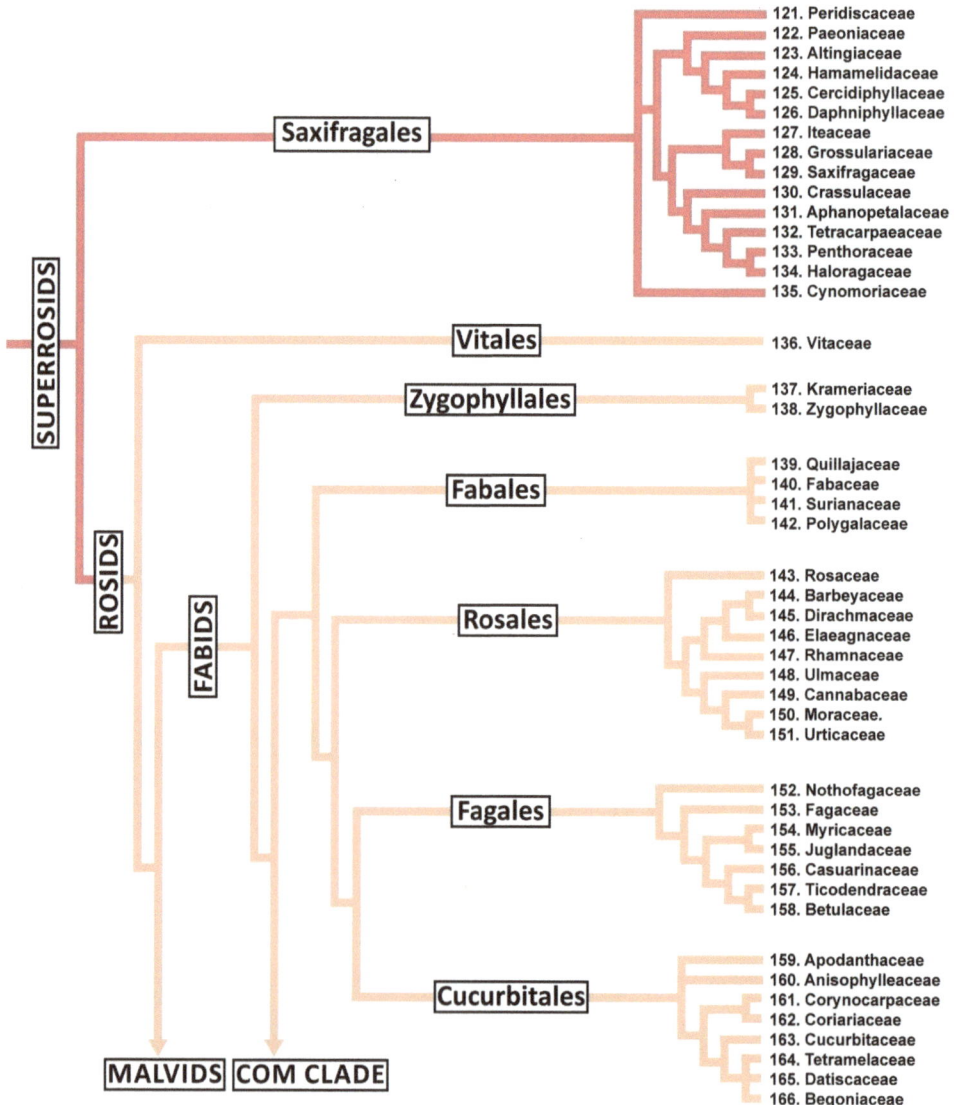

Saxifragales
- 121. Peridiscaceae
- 122. Paeoniaceae
- 123. Altingiaceae
- 124. Hamamelidaceae
- 125. Cercidiphyllaceae
- 126. Daphniphyllaceae
- 127. Iteaceae
- 128. Grossulariaceae
- 129. Saxifragaceae
- 130. Crassulaceae
- 131. Aphanopetalaceae
- 132. Tetracarpaeaceae
- 133. Penthoraceae
- 134. Haloragaceae
- 135. Cynomoriaceae

Vitales
- 136. Vitaceae

Zygophyllales
- 137. Krameriaceae
- 138. Zygophyllaceae

Fabales
- 139. Quillajaceae
- 140. Fabaceae
- 141. Surianaceae
- 142. Polygalaceae

Rosales
- 143. Rosaceae
- 144. Barbeyaceae
- 145. Dirachmaceae
- 146. Elaeagnaceae
- 147. Rhamnaceae
- 148. Ulmaceae
- 149. Cannabaceae
- 150. Moraceae.
- 151. Urticaceae

Fagales
- 152. Nothofagaceae
- 153. Fagaceae
- 154. Myricaceae
- 155. Juglandaceae
- 156. Casuarinaceae
- 157. Ticodendraceae
- 158. Betulaceae

Cucurbitales
- 159. Apodanthaceae
- 160. Anisophylleaceae
- 161. Corynocarpaceae
- 162. Coriariaceae
- 163. Cucurbitaceae
- 164. Tetramelaceae
- 165. Datiscaceae
- 166. Begoniaceae

SUPERROSIDS

ROSIDS

FABIDS

MALVIDS **COM CLADE**

Figure 12: Diagnostic characters and relationships of Saxifragales and rosids.

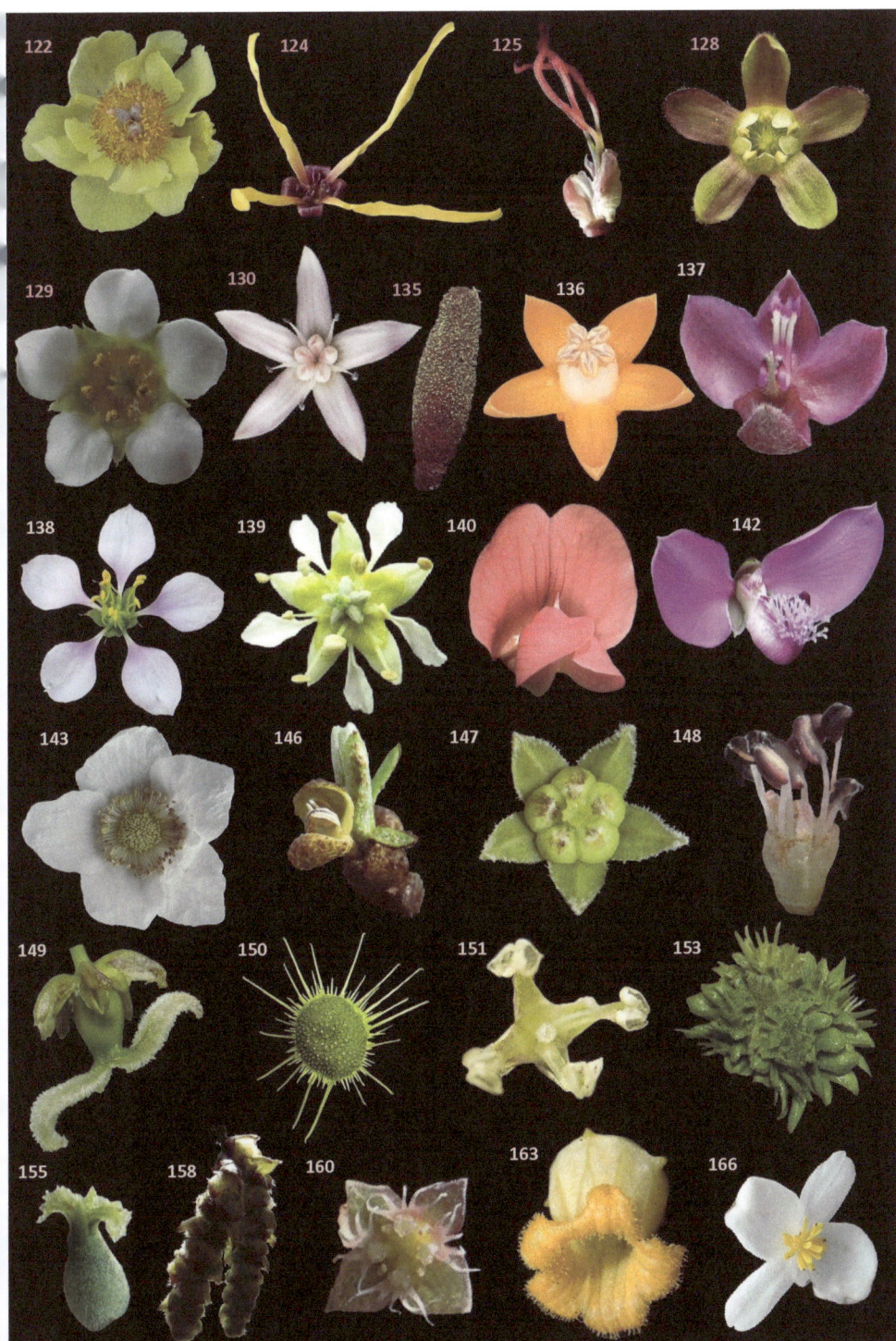

Figure 13: Floral images of Saxifragales (families 122-135), Vitales (136), Zygophyllales (137-138), Fabales (139-142), Rosales (143-151), Fagales (153-158) and Cucurbitales (160-166).

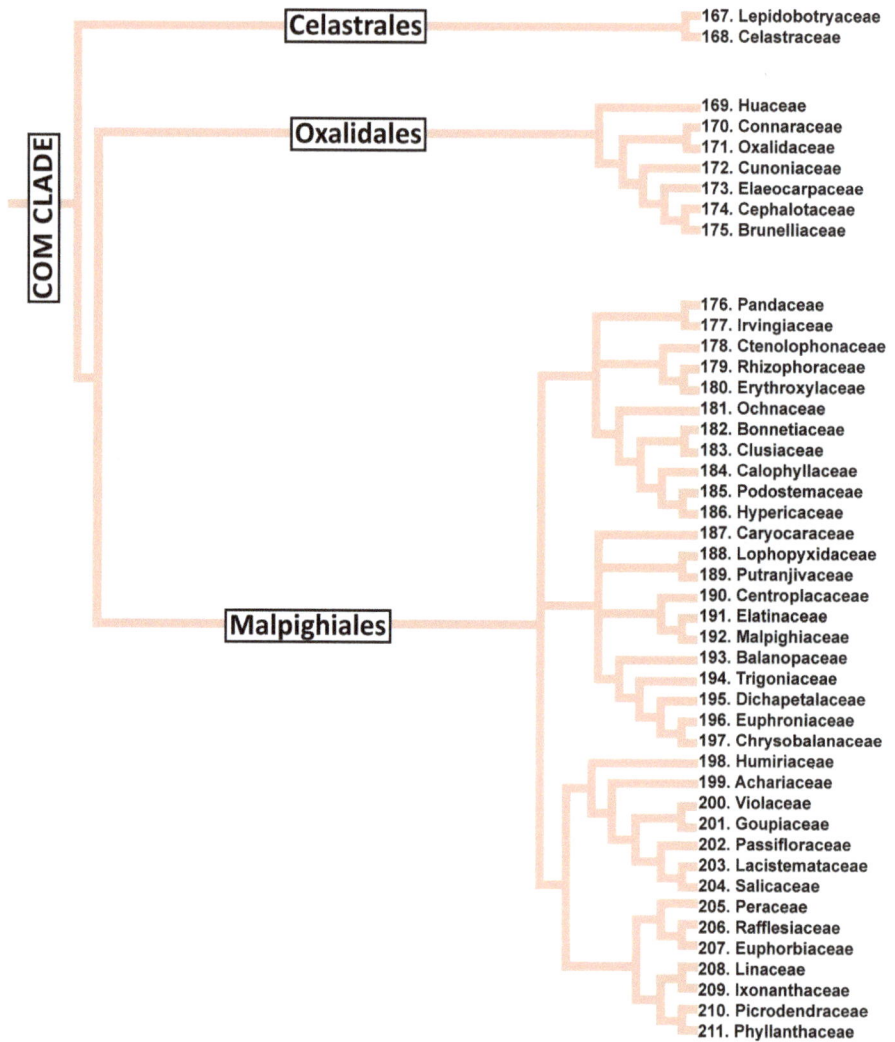

Figure 14: Relationships of the COM clade of rosids.

Figure 15: Floral images of Celastrales (family 168), Oxalidales (169-173) and Malpighiales (176-211).

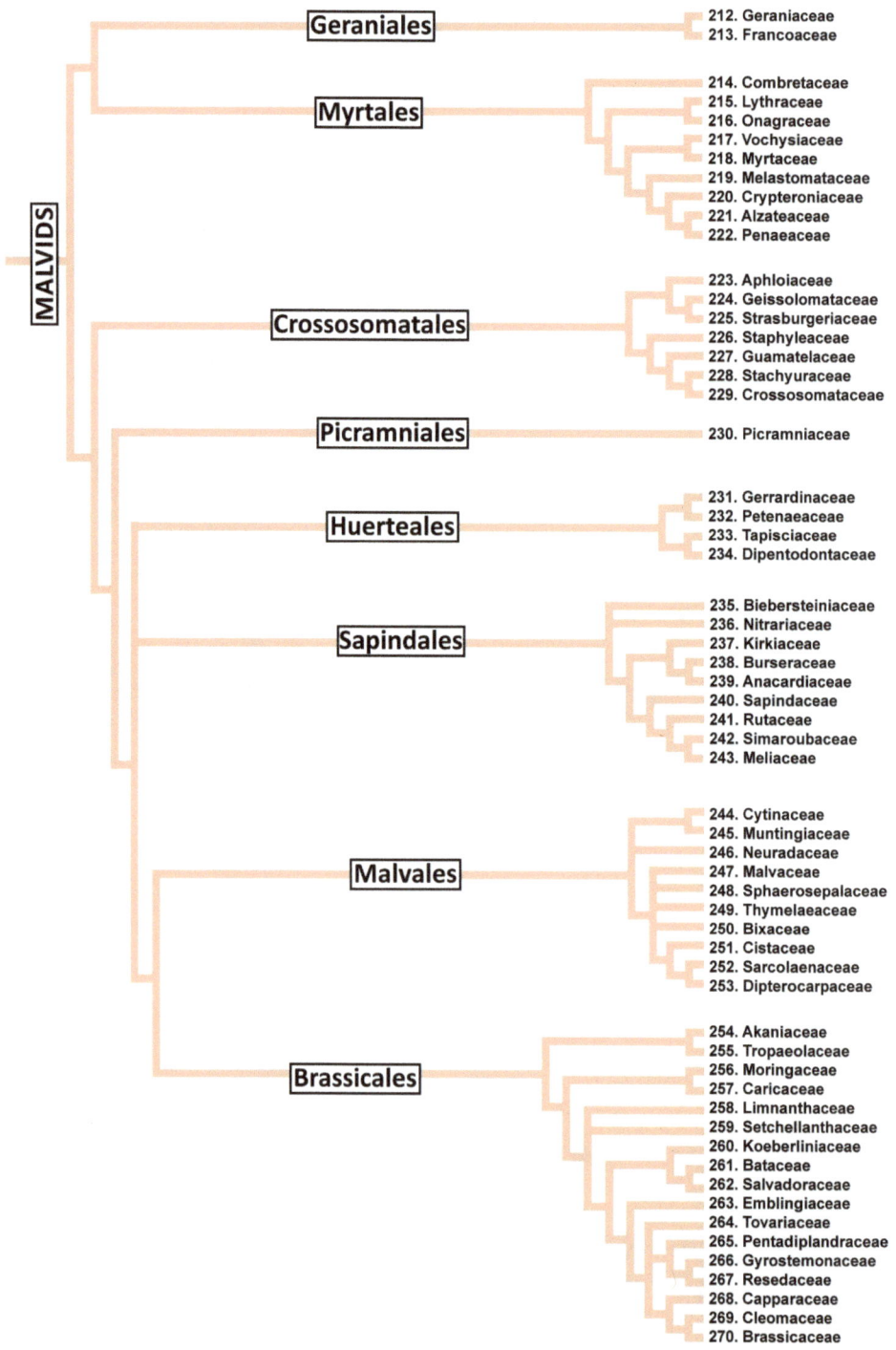

Figure 16: Relationships of malvid rosids.

The figure shows a phylogenetic tree labeled MALVIDS with the following orders and families:

Geraniales
- 212. Geraniaceae
- 213. Francoaceae

Myrtales
- 214. Combretaceae
- 215. Lythraceae
- 216. Onagraceae
- 217. Vochysiaceae
- 218. Myrtaceae
- 219. Melastomataceae
- 220. Crypteroniaceae
- 221. Alzateaceae
- 222. Penaeaceae

Crossosomatales
- 223. Aphloiaceae
- 224. Geissolomataceae
- 225. Strasburgeriaceae
- 226. Staphyleaceae
- 227. Guamatelaceae
- 228. Stachyuraceae
- 229. Crossosomataceae

Picramniales
- 230. Picramniaceae

Huerteales
- 231. Gerrardinaceae
- 232. Petenaeaceae
- 233. Tapisciaceae
- 234. Dipentodontaceae

Sapindales
- 235. Biebersteiniaceae
- 236. Nitrariaceae
- 237. Kirkiaceae
- 238. Burseraceae
- 239. Anacardiaceae
- 240. Sapindaceae
- 241. Rutaceae
- 242. Simaroubaceae
- 243. Meliaceae

Malvales
- 244. Cytinaceae
- 245. Muntingiaceae
- 246. Neuradaceae
- 247. Malvaceae
- 248. Sphaerosepalaceae
- 249. Thymelaeaceae
- 250. Bixaceae
- 251. Cistaceae
- 252. Sarcolaenaceae
- 253. Dipterocarpaceae

Brassicales
- 254. Akaniaceae
- 255. Tropaeolaceae
- 256. Moringaceae
- 257. Caricaceae
- 258. Limnanthaceae
- 259. Setchellanthaceae
- 260. Koeberliniaceae
- 261. Bataceae
- 262. Salvadoraceae
- 263. Emblingiaceae
- 264. Tovariaceae
- 265. Pentadiplandraceae
- 266. Gyrostemonaceae
- 267. Resedaceae
- 268. Capparaceae
- 269. Cleomaceae
- 270. Brassicaceae

Figure 17: Floral images of Geraniales (family 212), Myrtales (214-222), Crossosomatales (223-229), Huerteales (232-233) and Sapindales (237-241).

Figure 18: Floral images of Sapindales (families 242-243), Malvales (246-251) and Brassicales (255-270).

[Blank page]

Superasterids

Berberidopsidales

Stomata cyclocytic

Stipules absent

Perianth free

Stamen filaments stout

Ovaries superior

Fruits fleshy

Santalales

Plants predominantly parasitic

Sepals often reduced, known as a 'calyculus'

Caryophyllales

Petals absent, if present likely of staminal origin

Adaptations to 'stressful' environments common (e.g. arid, nitrogen poor soils, halophytic)

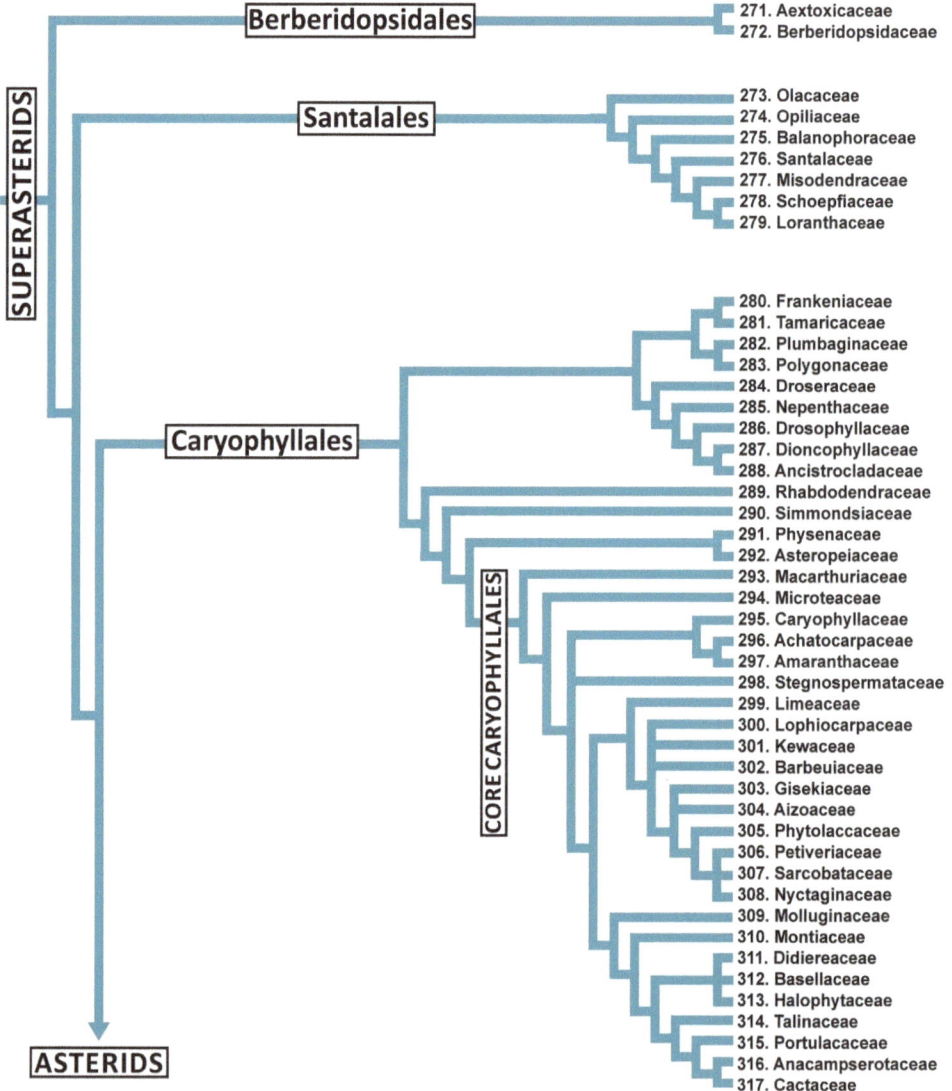

Figure 19: Diagnostic characters and relationships of Berberidopsidales, Santalales and Caryophyllales.

24

Figure 20: Floral images of Berberidopsidales (family 272), Santalales (273-279) and Caryophyllales (280-317).

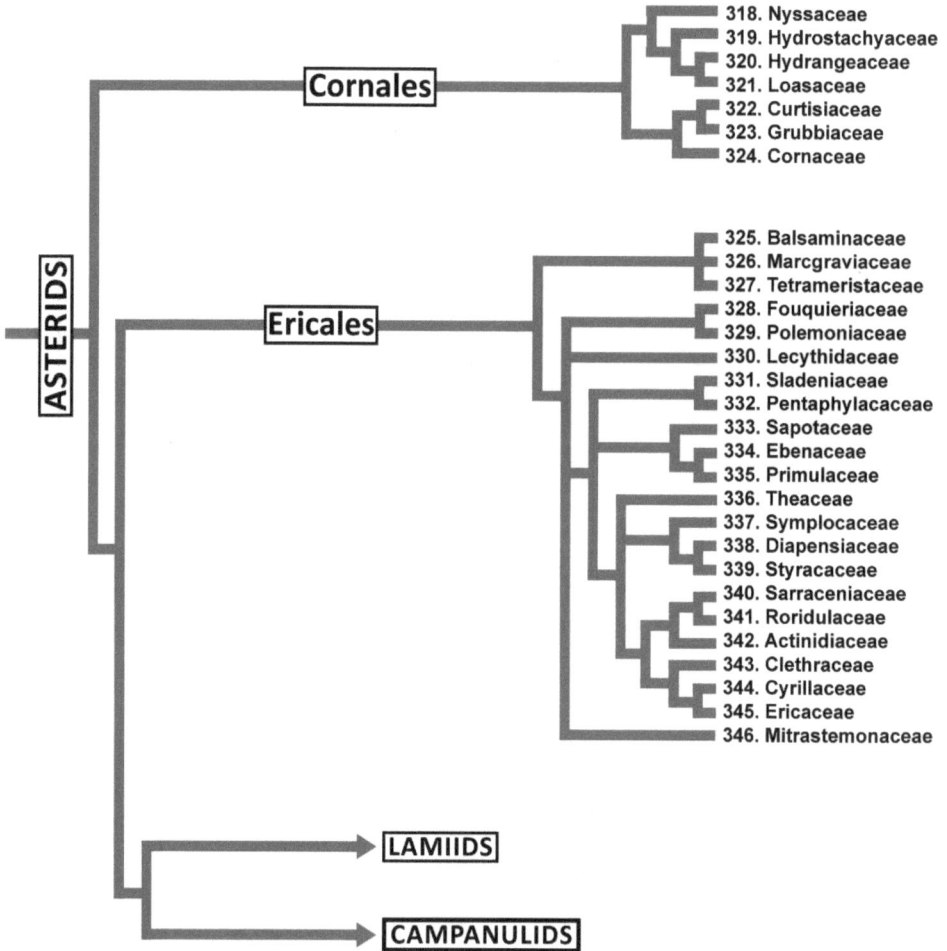

Figure 21: Diagnostic characters and relationships of asterids.

Figure 22: Floral images of Cornales (families 319-324) and Ericales (325-346).

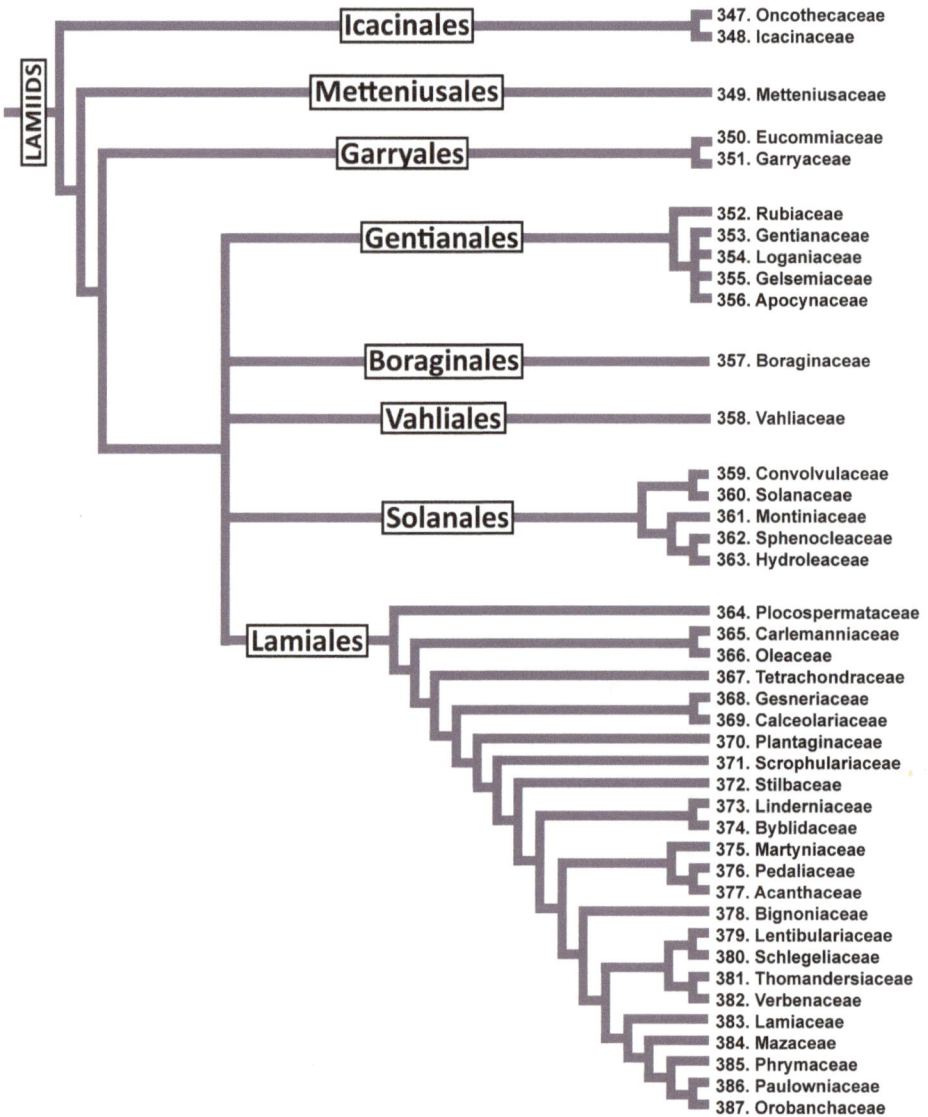

Figure 23: Relationships of lamiid asterids.

Icacinales	347. Oncothecaceae
	348. Icacinaceae
Metteniusales	349. Metteniusaceae
Garryales	350. Eucommiaceae
	351. Garryaceae
Gentianales	352. Rubiaceae
	353. Gentianaceae
	354. Loganiaceae
	355. Gelsemiaceae
	356. Apocynaceae
Boraginales	357. Boraginaceae
Vahliales	358. Vahliaceae
Solanales	359. Convolvulaceae
	360. Solanaceae
	361. Montiniaceae
	362. Sphenocleaceae
	363. Hydroleaceae
Lamiales	364. Plocospermataceae
	365. Carlemanniaceae
	366. Oleaceae
	367. Tetrachondraceae
	368. Gesneriaceae
	369. Calceolariaceae
	370. Plantaginaceae
	371. Scrophulariaceae
	372. Stilbaceae
	373. Linderniaceae
	374. Byblidaceae
	375. Martyniaceae
	376. Pedaliaceae
	377. Acanthaceae
	378. Bignoniaceae
	379. Lentibulariaceae
	380. Schlegeliaceae
	381. Thomandersiaceae
	382. Verbenaceae
	383. Lamiaceae
	384. Mazaceae
	385. Phrymaceae
	386. Paulowniaceae
	387. Orobanchaceae

Figure 24: Floral images of Icacinales (family 348), Metteniusales (349), Garryales (351) and Gentianales (352).

28

Figure 25: Floral images of Gentianales (families 353-356), Boraginales (357), Solanales (359-360) and Lamiales (366-387).

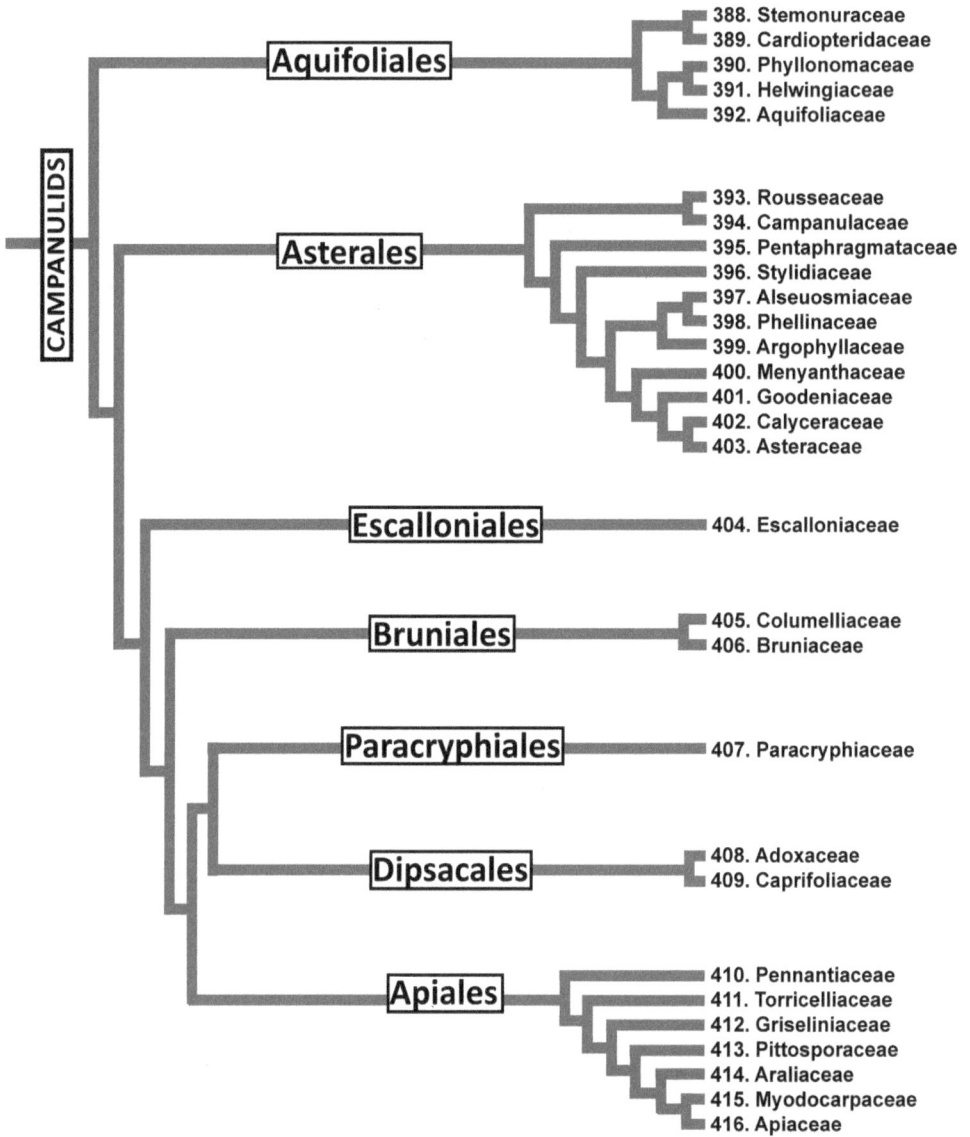

Figure 26: Relationships of campanulid asterids.

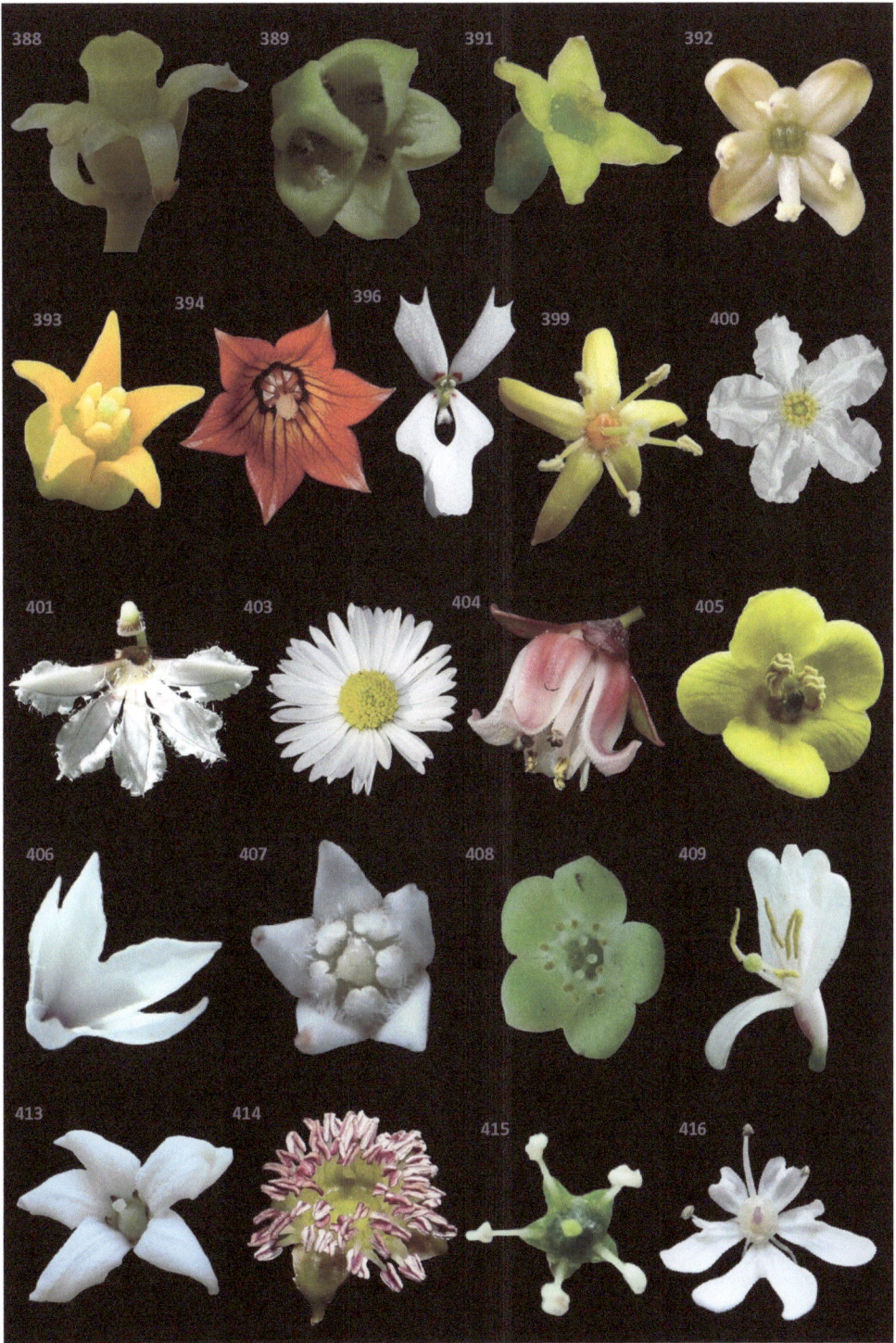

Figure 27: Floral images of Aquifoliales (families 388-392), Asterales (393-403), Escalloniales (404), Bruniales (405-406), Paracryphiales (407), Dipsacales (408-409) and Apiales (413-416).

Appendix 2: List of images

1. *Amborella trichopoda* Baill. (Amborellaceae)
   © Mike Bayly (adapted: CC BY-SA 3.0)
2. *Trithuria filamentosa* Rodway (Hydatellaceae)
   © Chris Davidson
3. *Cabomba caroliniana* A.Gray (Cabombaceae)
   © Rogier van Vugt
4. *Nymphaea caerulea* Savigny (Nymphaeaceae)
   © Rogier van Vugt
5. *Austrobaileya scandens* C.T.White (Austrobaileyaceae)
   © Chris Davidson
7. *Illicium henryi* Diels (Schisandraceae)
   © Scott Zona (adapted: CC BY 2.0)
9. *Tasmannia piperita* (Hook.f.) Miers (Winteraceae)
   © Rogier van Vugt
10. *Anemopsis californica* (Nutt.) Hook. & Arn. (Saururaceae)
    © James Byng
11. *Piper nigrum* L. (Piperaceae)
    © James Byng
12. *Aristolochia macrophylla* Lam. (Aristolochiaceae)
    © Maarten Christenhusz
13. *Scyphocephalium mannii* (Benth. & Hook. f.) Warb. (Myristicaceae) © Ehoarn Bidault
14. *Magnolia sprengeri* Pamp. (Magnoliaceae)
    © James Byng
15. *Degeneria vitiensis* L.W.Bailey & A.C.Sm. (Degeneriaceae)
    © Chris Davidson
17. *Eupomatia laurina* R.Br. (Eupomatiaceae)
    © Rogier van Vugt
18. *Annona reticulata* L. (Annonaceae)
    © Rogier van Vugt
19. *Calycanthus occidentalis* Hook. & Arn. (Calycanthaceae)
    © James Byng
22. *Laurelia sempervirens* (Ruiz & Pav.) Tul. (Atherospermataceae) © James Byng
23. *Hernandia nymphaeifolia* (C.Presl) Kubitzki (Hernandiaceae)
    © Rogier van Vugt
24. *Tambourissa* sp. (Monimiaceae)
    © James Byng
25. *Apollonias barbujana* (Cav.) Bornm. (Lauraceae)
    © Rogier van Vugt
26. *Chloranthus spicatus* (Thunb.) Makino (Chloranthaceae)
    © James Byng
27. *Acorus calamus* L. (Acoraceae)
    © James Byng
28. *Amorphophallus titanum* (Becc.) Becc. ex Arcang. (Araceae) © Rogier van Vugt
29. *Harperocallis flava* McDaniel (Tofieldiaceae)
    © Walter Siegmund (adapted: CC BY-SA 3.0)
30. *Echinodorus paniculatus* Micheli (Alismataceae)
    © James Byng
31. *Butomus umbellatus* L. (Butomaceae)
    © James Byng
32. *Hydrocharis morsus-ranae* L. (Hydrocharitaceae)
    © Rogier van Vugt
34. *Aponogeton distachyos* L.f. (Aponogetonaceae)
    © James Byng
35. *Triglochin palustris* L. (Juncaginaceae)
    © Rogier van Vugt
38. *Potamogeton gramineus* L. (Potamogetonaceae)
    © Pellaea (adapted: CC BY 2.0)
42. *Petrosavia stellaris* Becc. (Petrosaviaceae)
    © Rogier van Vugt
44. *Gymnosiphon bekensis* Letouzey (Burmanniaceae)
    © Vincent Merckx

45. *Dioscorea* sp. (Dioscoreaceae)
    © Maarten Christenhusz
46. *Sciaphila densiflora* Schltr. (Triuridaceae)
    © Ehoarn Bidault
47. *Xerophyta elegans* (Balf.) Baker (Velloziaceae)
    © James Byng
48. *Croomia heterosepala* (Baker) Okuyama (Stemonaceae)
    © Keisotyo (adapted: CC BY-SA 3.0)
50. *Freycinetia beccarii* Solms (Pandanaceae)
    © Fritz Geller-Grimm (adapted: CC BY-SA 3.0)
52. *Corsia ornata* Becc. (Corsiaceae)
    © Thassilo Franke (adapted: CC BY-SA 3.0)
53. *Trillium flexipes* Raf. (Melanthiaceae)
    © James Byng
55. *Alstroemeria pallida* Graham (Alstroemeriaceae)
    © James Byng
56. *Colchicum speciosum* Steven (Colchicaceae)
    © James Byng
57. *Lapageria rosea* Ruiz & Pav. (Philesiaceae)
    © James Byng
59. *Smilax china* L. (Smilacaceae)
    © James Byng
60. *Lilium regale* E.H.Wilson (Liliaceae)
    © James Byng
61. *Ophrys scolopax* Cav. (Orchidaceae)
    © Rogier van Vugt
63. *Blandfordia nobilis* Sm. (Blandfordiaceae)
    © Peter Woodard (CCO 1.0)
66. *Hypoxis hirsuta* (L.) Coville (Hypoxidaceae)
    © James Byng
67. *Doryanthes palmeri* W.Hill ex Benth. (Doryanthaceae)
    © James Byng
68. *Ixiolirion tataricum* (Pall.) Herb. (Ixioliriaceae)
    © C.T. Johansson (adapted: CC BY-SA 3.0)
69. *Tecophilaea cyanocrocus* Leyb. (Tecophilaeaceae)
    © James Byng
70. *Crocus nudiflorus* Sm. (Iridaceae)
    © Maarten Christenhusz
72. *Hemerocallis* 'Stella de Oro' (Asphodelaceae)
    © James Byng
73. *Narcissus pseudonarcissus* L. (Amaryllidaceae)
    © Rogier van Vugt
74. *Ornithogalum thyrsoides* Jacq. (Asparagaceae)
    © Rogier van Vugt
76. *Hyophorbe lagenicaulis* (L.H.Bailey) H.E.Moore (Arecaceae) © James Byng
78. *Commelina communis* L. (Commelinaceae)
    © James Byng
79. *Helmholtzia glaberrima* (Hook.f.) Caruel (Philydraceae)
    © Stan Shebs (adapted: CC BY-SA 3.0)
80. *Eichhornia crassipes* (Mart.) Solms (Pontederiaceae)
    © James Byng
81. *Wachendorfia paniculata* Burm. (Haemodoraceae)
    © Rogier van Vugt
82. *Strelitzia reginae* Aiton (Strelitziaceae)
    © Maarten Christenhusz
84. *Heliconia aemygdiana* Burle-Marx (Heliconiaceae)
    © Chris Davidson
85. *Musa beccarii* N.W.Simmonds (Musaceae)
    © Chris Davidson
86. *Canna indica* L. (Cannaceae)
    © James Byng
87. *Maranta leuconeura* E.Morren (Marantaceae)
    © Maarten Christenhusz
88. *Hellenia speciosa* (J.Koenig) Govaerts (Costaceae)
    © James Byng
89. *Alpinia havilandii* K.Schum. (Zingiberaceae)
    © Rogier van Vugt

90. *Sparganium natans* L. (Typhaceae)
© Rogier van Vugt
91. *Vriesea heterostachys* (Baker) L.B.Sm. (Bromeliaceae)
© James Byng
93. *Xyris* sp. (Xyridaceae)
© James Byng
94. *Eriocaulon compressum* Lam. (Eriocaulaceae)
© Bob Peterson (adapted: CC BY 2.0)
95. *Mayaca longipes* Mart. ex Seub. (Mayacaceae)
© Chris Davidson
96. *Thurnia sphaerocephala* (Rudge) Hook.f. (Thurniaceae)
© Chris Davidson
97. *Luzula campestris* (L.) DC. (Juncaceae)
© Rogier van Vugt
98. *Carex flacca* Schreb. (Cyperaceae)
© Rogier van Vugt
99. *Sporadanthus* sp. (Restionaceae)
© Kevin Thiele (adapted: CC BY 2.0)
102. *Georgeantha hexandra* B.G.Briggs & L.A.S.Johnson
(Ecdeiocoleaceae) © Maarten Christenhusz
103. *Phleum pratense* L. (Poaceae)
© Rogier van Vugt
106. *Papaver rhoeas* L. (Papaveraceae)
© James Byng
107. *Kingdonia uniflora* Balf.f. & W.W.Sm. (Circaeasteraceae)
© Chris Davidson
108. *Akebia quinata* (Houtt.) Decne. (Lardizabalaceae)
© Alpsdake (adapted: CC-BY-SA-3.0)
109. *Stephania glandulifera* Miers (Menispermaceae)
© James Byng
110. *Epimedium stellulatum* Stearn (Berberidaceae)
© James Byng
111. *Helleborus orientalis* Lam. (Ranunculaceae)
© James Byng
112. *Sabia yunnanensis* Franch. (Sabiaceae)
© Chris Davidson
113. *Nelumbo nucifera* Gaertn. (Nelumbonaceae)
© James Byng
114. *Platanus ×hispanica* Mill. ex Münchh. (Platanaceae)
© James Byng
115. *Grevillea plurijuga* F.Muell. (Proteaceae)
© Maarten Christenhusz
116. *Trochodendron aralioides* Siebold & Zucc. (Trochodend-
raceae) © Qwert1234
117. *Sarcococca confusa* Sealy (Buxaceae)
© James Byng
120. *Dillenia suffruticosa* (Griff.) Martelli (Dilleniaceae)
© James Byng
122. *Paeonia daurica* Andrews subsp. *mlokosewitschii* (Loma-
kin) D.Y.Hong (Paeoniaceae) © James Byng
124. *Hamamelis ×intermedia* Rehder (Hamamelidaceae)
© James Byng
125. *Cercidiphyllum japonicum* Siebold & Zucc. ex J.J.Hoffm.
& J.H.Schult.bis (Cercidiphyllaceae) © James Byng
128. *Ribes uva-crispa* L. (Grossulariaceae)
© Frank Vincentz (adapted: CC-BY-SA-3.0)
129. *Heuchera villosa* Michx. (Saxifragaceae)
© James Byng
130. *Crassula ovata* (Mill.) Druce (Crassulaceae)
© James Byng
135. *Cynomorium coccineum* L. (Cynomoriaceae)
© Chris Davidson
136. *Leea guineensis* G.Don (Vitaceae)
© Ehoarn Bidault
137. *Krameria tomentosa* A.St.-Hil. (Krameriaceae)
© Joao Medeiros (adapted: CC BY 2.0)
138. *Fagonia cretica* L. (Zygophyllaceae)
© Llez (adapted: CC-BY-SA-3.0)

139. *Quillaja saponaria* Molina (Quillajaceae)
© Chris Davidson
140. *Lathyrus nissolia* L. (Fabaceae)
© Rogier van Vugt
142. *Polygala myrtifolia* L. (Polygalaceae)
© James Byng
143. *Rubus deliciosus* Torr. (Rosaceae)
© James Byng
146. *Hippophaë rhamnoides* L. (Elaeagnaceae)
© James Byng
147. *Maesopsis eminii* Engl. (Rhamnaceae)
© Ehoarn Bidault
148. *Ulmus glabra* Huds. (Ulmaceae)
© Hermann Schachner (adapted: CC0)
149. *Celtis occidentalis* L. (Cannabaceae)
© James Byng
150. *Dorstenia christenhuszii* M.W.Chase & M.F.Fay (Moraceae)© Maarten Christenhusz
151. *Urtica dioica* L. (Urticaceae)
© Frank Vincentz (adapted: CC-BY-SA-3.0)
153. *Castanea sativa* Mill. (Fagaceae)
© Rogier van Vugt
155. *Juglans regia* L. (Juglandaceae)
© N-Baudet (adapted: CC-BY-SA-3.0)
158. *Carpinus betulus* L. (Betulaceae)
© James Byng
160. *Anisophyllea disticha* (Jack) Baill. (Anisophylleaceae)
© Chris Davidson
163. *Coccinia racemiflora* Keraudren (Cucurbitaceae)
© Ehoarn Bidault
166. *Begonia minor* Jacq. (Begoniaceae)
© James Byng
168. *Putterlickia pyracantha* (L.) Endl. (Celastraceae)
© James Byng
169. *Afrostyrax kamerunensis* Perkins & Gilg (Huaceae)
© Ehoarn Bidault
170. *Manotes griffoniana* Baill. (Connaraceae)
© Ehoarn Bidault
171. *Oxalis acetosella* L. (Oxalidaceae)
© James Byng
173. *Crinodendron hookerianum* Gay (Elaeocarpaceae)
© Maarten Christenhusz
176. *Microdesmis keayana* J.Léonard (Pandaceae)
© Ehoarn Bidault
179. *Bruguiera gymnorhiza* (L.) Savigny (Rhizophoraceae)
© James Byng
180. *Erythroxylum sechellarum* O.E.Schulz (Erythroxylaceae)
© Rogier van Vugt
181. *Ochna thomasiana* Engl. & Gilg (Ochnaceae)
© James Byng
183. *Clusia rosea* Jacq. (Clusiaceae)
© James Byng
186. *Hypericum lanceolatum* Lam. (Hypericaceae)
© Rogier van Vugt
189. *Drypetes* sp. (Putranjivaceae)
© Ehoarn Bidault
192. *Malpighia coccigera* L. (Malpighiaceae)
© Sten Porse (adapted: CC-BY-SA-3.0)
195. *Dichapetalum heudelotii* (Planch ex Oliv.) Baill. var.
hispidum (Oliv.) Breteler (Dichapetalaceae)
© Ehoarn Bidault
197. *Chrysobalanus icaco* L. (Chrysobalanaceae)
© Rogier van Vugt
199. *Hydnocarpus pentandrus* (Buch.-Ham.) Oken (Acharia-
ceae) © Jayesh Patil 912 (adapted: CC BY 2.0)
200. *Viola delavayi* Franch. (Violaceae)
© James Byng
202. *Passiflora incarnata* L. (Passifloraceae)
© James Byng

204. *Dovyalis zenkeri* Gilg (Salicaceae)
© Ehoarn Bidault
205. *Clutia pulchella* L. (Peraceae)
© Rogier van Vugt
206. *Rafflesia keithii* Meijer (Rafflesiaceae)
© Rogier van Vugt
207. *Euphorbia paralias* L. (Euphorbiaceae)
© Rogier van Vugt
208. *Linum suffruticosum* L. (Linaceae)
© James Byng
211. *Maesobotrya klaineana* (Pierre) J.Léonard (Phyllanthaceae) © Ehoarn Bidault
212. *Pelargonium echinatum* Curtis (Geraniaceae)
© James Byng
214. *Strephonema mannii* Hook.f. (Combretaceae)
© Ehoarn Bidault
215. *Cuphea lanceolata* W.T.Aiton (Lythraceae)
© Maarten Christenhusz
216. *Epilobium hirsutum* L. (Onagraceae)
© Rogier van Vugt
218. *Eucalyptus debeuzevillei* Maiden (Myrtaceae)
© Rogier van Vugt
219. *Osbeckia porteresii* Jacq.-Fél. (Melastomataceae)
© Ehoarn Bidault
221. *Alzatea verticillata* Ruiz & Pav. (Alzateaceae)
© Paul Maas
222. *Saltera sarcocolla* Bullock (Penaeaceae)
© Rogier van Vugt
223. *Aphloia theiformis* (Vahl) Benn. (Aphloiaceae)
© Rogier van Vugt
226. *Staphylea bolanderi* A.Gray (Staphyleaceae)
© Stan Shebs (adapted: CC-BY-SA-3.0)
227. *Guamatela tuerckheimii* Donn.Sm.(Guamatelaceae)
© Chris Davidson
228. *Stachyurus chinensis* Franch. (Stachyuraceae)
© James Byng
229. *Crossosoma californicum* Nutt. (Crossosomataceae)
© John Game (adapted: CC BY 2.0)
232. *Petenaea cordata* Lundell (Petenaeaceae)
© Chris Davidson
233. *Tapiscia sinensis* Oliv. (Tapisciaceae)
© Maarten Christenhusz
237. *Kirkia acuminata* Oliv. (Kirkiaceae)
© James Byng
238. *Dacryodes buettneri* (Engl.) H.J.Lam (Burseraceae)
© Ehoarn Bidault
239. *Rhus typhina* L. (Anacardiaceae)
© Maarten Christenhusz
240. *Allophylus edulis* (A.St.-Hil., A.Juss. & Cambess.) Hieron. ex Niederl. (Sapindaceae) © Mauricio Bonifacino
241. *Citrus sp.* (Rutaceae)
© James Byng
242. *Quassia africana* (Baill.) Baill. (Simaroubaceae)
© Ehoarn Bidault
243. *Turraea heterophylla* Sm. (Meliaceae)
© James Byng
246. *Grielum grandiflorum* (L.) Druce (Neuradaceae)
© Andrew Massyn (adapted: CC-BY-SA-3.0)
247. *Hibiscus fragilis* DC. (Malvaceae)
© James Byng
249. *Daphne odora* Thunb. (Thymelaeaceae)
© James Byng
250. *Cochlospermum religiosum* (L.) Alston (Bixaceae)
© Prenn (adapted: CC-BY-SA-3.0)
251. *Helianthemum nummularium* Mill. (Cistaceae)
© James Byng
255. *Tropaeolum majus* L. (Tropaeolaceae)
© James Byng

256. *Moringa oleifera* Lam. (Moringaceae)
© Alexey Sergeev
257. *Carica papaya* L. (Caricaceae)
© H.Zell (adapted: CC-BY-SA-3.0)
258. *Limnanthes douglasii* R.Br. (Limnanthaceae)
© James Byng
265. *Pentadiplandra brazzeana* Baill. (Pentadiplandraceae)
© Ehoarn Bidault
267. *Reseda luteola* L. (Resedaceae)
© Rogier van Vugt
268. *Capparis lasiantha* R.Br. ex DC. (Capparaceae)
© Mark Marathon (adapted: CC-BY-SA-3.0)
269. *Cleome maculata* (Sond.) Szyszył. (Cleomaceae)
© Rogier van Vugt
270. *Arabis cypria* Holmboe (Brassicaceae)
© James Byng
272. *Berberidopsis corallina* Hook.f. (Berberidopsidaceae)
© James Byng
273. *Olax obtusifolia* De Wild. (Olacaceae)
© James Byng
275. *Thonningia sanguinea* Vahl (Balanophoraceae)
© Ehoarn Bidault
276. *Thesium subsucculentum* (Kämmer) J.C.Manning & F.Forest (Santalaceae) © Rogier van Vugt
279. *Helixanthera mannii* (Oliv.) Danser (Loranthaceae)
© Ehoarn Bidault
280. *Frankenia ericifolia* C.Sm. ex DC. (Frankeniaceae)
© Rogier van Vugt
281. *Tamarix aucheriana* (Decne. ex Walp.) B.R.Baum (Tamaricaceae) © Alexey Sergeev
282. *Plumbago auriculata* Lam. (Plumbaginaceae)
© James Byng
283. *Coccoloba uvifera* (L.) L. (Polygonaceae)
© Maarten Christenhusz
284. *Dionaea muscipula* J.Ellis (Droseraceae)
© James Byng
285. *Nepenthes pervillei* Blume (Nepenthaceae)
© Rogier van Vugt
286. *Drosophyllum lusitanicum* (L.) Link (Drosophyllaceae)
© Javier Martin
295. *Stellaria holostea* L. (Caryophyllaceae)
© Maarten Christenhusz
297. *Maireana turbinata* Paul G.Wilson (Amaranthaceae)
© Maarten Christenhusz
299. *Limeum africanum* L. (Limeaceae)
© Chris Davidson
301. *Kewa bowkeriana* (Sond.) Christenh. (Kewaceae)
© Maarten Christenhusz
302. *Barbeuia madagascariensis* Steud. (Barbeuiaceae)
© Chris Davidson
304. *Mesembryanthemum sp.* (Aizoaceae)
© Rogier van Vugt
308. *Mirabilis longiflora* L. (Nyctaginaceae)
© Maarten Christenhusz
310. *Lewisia cotyledon* (S.Watson) B.L.Rob. (Montiaceae)
© James Byng
312. *Anredera cordifolia* (Ten.) Steenis (Basellaceae)
© Andrew Massyn
314. *Talinum napiforme* DC. (Talinaceae)
© Michael Wolf (adapted: CC-BY-SA-3.0)
315. *Portulaca grandiflora* Hook. (Portulacaceae)
© James Byng
316. *Anacampseros crinita* Dinter (Anacampserotaceae)
© James Byng
317. *Opuntia sp.* (Cactaceae)
© James Byng
318. *Davidia involucrata* Baill. (Nyssaceae)
© James Byng

320. *Philadelphus microphyllus* A.Gray (Hydrangeaceae)
© James Byng

321. *Nasa triphylla* (Juss.) Weigend (Loasaceae)
© H.Zell (adapted: CC-BY-SA-3.0)

323. *Grubbia rosmarinifolia* P.J.Bergius (Grubbiaceae)
© Chris Davidson

324. *Cornus kousa* Bürger ex Hance (Cornaceae)
© James Byng

325. *Impatiens pseudoviola* Gilg (Balsaminaceae)
© Maarten Christenhusz

326. *Schwartzia costaricensis* (Gilg) Bedell (Marcgraviaceae)
© Chris Davidson

328. *Fouquieria diguetii* (Tiegh.) I.M.Johnst. (Fouquieriaceae)
© Maarten Christenhusz

329. *Cobaea scandens* Cav. (Polemoniaceae)
© James Byng

330. *Couroupita guianensis* Aubl. (Lecythidaceae)
© James Byng

332. *Visnea mocanera* L.f. (Pentaphylacaceae)
© James Byng

333. *Pouteria* sp. (Sapotaceae)
© Ehoarn Bidault

334. *Diospyros boiviniana* (Baill.) G.E.Schatz & Lowry (Ebenaceae) © Rogier van Vugt

335. *Cyclamen pseudibericum* Hildebr. (Primulaceae)
© Rogier van Vugt

336. *Camellia saluenensis* Stapf ex Bean (Theaceae)
© James Byng

339. *Sinojackia rehderiana* H.H.Hu (Styracaceae)
© James Byng

340. *Sarracenia purpurea* L. (Sarraceniaceae)
© James Byng

341. *Roridula gorgonias* Planch. (Roridulaceae)
© James Byng

342. *Actinidia chinensis* Planch. (Actinidiaceae)
© Chris Davidson

345. *Pyrola rotundifolia* L. (Ericaceae)
© Rogier van Vugt

346. *Mitrastemma yamamotoi* Makino (Mitrastemonaceae)
© Rogier van Vugt

348. *Pyrenacantha malvifolia* Engl. (Icacinaceae)
© Maarten Christenhusz

349. *Rhaphiostylis beninensis* (Hook.f. ex Planch.) Planch. ex Benth. (Metteniusaceae) © Ehoarn Bidault

351. *Garrya elliptica* Douglas ex Lindl. (Garryaceae)
© James Byng

352. *Rothmannia annae* (E.P.Wright) Keay (Rubiaceae)
© Rogier van Vugt

353. *Gentiana paradoxa* Albov (Gentianaceae)
© James Byng

355. *Mostuea megaphylla* R.D.Good (Gelsemiaceae)
© Ehoarn Bidault

356. *Vinca minor* L. (Apocynaceae)
© Rogier van Vugt

357. *Omphalodes verna* Moench (Boraginaceae)
© Maarten Christenhusz

359. *Convolvulus floridus* L.f. (Convolvulaceae)
© James Byng

360. *Solanum americanum* Mill. (Solanaceae)
© James Byng

366. *Forsythia suspensa* (Thunb.) Vahl (Oleaceae)
© James Byng

368. *Streptocarpus ionanthus* (H.Wendl.) Christenh. (Gesneriaceae) © Rogier van Vugt

369. *Calceolaria biflora* Lam. (Calceolariaceae)
© James Byng

370. *Plantago lanceolata* L. (Plantaginaceae)
© Rogier van Vugt

371. *Verbascum phoeniceum* L. (Scrophulariaceae)
© Maarten Christenhusz

372. *Halleria lucida* L. (Stilbaceae)
© Abu Shawka

374. *Byblis liniflora* Salisb. (Byblidaceae)
© Rogier van Vugt

377. *Thunbergia alata* Bojer ex Sims (Acanthaceae)
© James Byng

378. *Catalpa bignonioides* Walter (Bignoniaceae)
© James Byng

379. *Pinguicula moranensis* Kunth (Lentibulariaceae)
© James Byng

382. *Verbena officinalis* L. (Verbenaceae)
© Rogier van Vugt

383. *Lamium maculatum* L. (Lamiaceae)
© Rogier van Vugt

384. *Mazus pumilus* (Burm.f.) Steenis (Mazaceae)
© Alpsdake (adapted: CC-BY-SA-3.0)

385. *Erythranthe naiandina* (J.M.Watson & C.Bohlen) G.L.Nesom (Phrymaceae) © James Byng

386. *Paulownia tomentosa* (Thunb.) Steud. (Paulowniaceae)
© James Byng

387. *Orobanche purpurea* Jacq. (Orobanchaceae)
© Rogier van Vugt

388. *Gomphandra mappioides* Valeton (Stemonuraceae)
© James Byng

389. *Gonocaryum litorale* (Blume) Sleumer (Cardiopteridaceae) © James Byng

391. *Helwingia chinensis* Batalin (Helwingiaceae)
© Rogier van Vugt

392. *Ilex integra* Thunb. (Aquifoliaceae)
© Maarten Christenhusz

393. *Roussea simplex* Sm. (Rousseaceae)
© Vincent Florens

394. *Canarina canariensis* (L.) Vatke (Campanulaceae)
© Rogier van Vugt

396. *Stylidium androsaceum* Lindl. (Stylidiaceae)
© Maarten Christenhusz

399. *Corokia macrocarpa* Kirk (Argophyllaceae)
© James Byng

400. *Nymphoides ezannoi* Berhaut (Menyanthaceae)
© James Byng

401. *Scaevola taccada* (Gaertn.) Roxb. (Goodeniaceae)
© Rogier van Vugt

403. *Bellis perennis* L. (Asteraceae)
© Rogier van Vugt

404. *Forgesia racemosa* J.F.Gmel. (Escalloniaceae)
© Rogier van Vugt

405. *Columellia oblonga* Ruiz & Pav. (Columelliaceae)
© Chris Davidson

406. *Brunia monogyna* (Vahl) Class.-Bockh. & E.G.H.Oliv. (Bruniaceae) © Chris Davidson

407. *Quintinia quatrefagesii* F.Muell. (Paracryphiaceae)
© Chris Davidson

408. *Adoxa moschatellina* L. (Adoxaceae)
© James Byng

409. *Lonicera xylosteum* L. (Caprifoliaceae)
© Maarten Christenhusz

413. *Pittosporum coriaceum* Dryand. ex Aiton (Pittosporaceae) © James Byng

414. *Polyscias crassa* (Hemsl.) Lowry & G.M.Plunkett (Araliaceae) © Rogier van Vugt

415. *Myodocarpus crassifolius* Dubard & R.Vig. (Myodocarpaceae) © Chris Davidson

416. *Anthriscus sylvestris* (L.) Hoffm. (Apiaceae)
© James Byng